Aerial Robotics

With

STM32F100RB

Microcontroller

S. M. Ibraheem

The University of Lahore

Copyright Notice:

Copyright © 2021. All rights are reserved.

This book is self-published by the author

This book is copy righted through creative commons

Aerial Robotics with STM32F100RB by Sheikh Muhammad Ibraheem is licensed under a <u>Creative Commons Attribution-Noncommercial 4.0 International License</u>.

Dedication

I dedicate this book to my parents. Who supported me in every moment in my life and to the teachers who taught me the knowledge of the world.

Acknowledgement

I would like to thank my supervisor Engr. Arslan Shahid, for his guidance during the project

I would like to acknowledge my professors, mentors and friends who encouraged me to do this.

Sheikh Muhammad Ibraheem

About Author:

Sheikh Muhammad Ibraheem born in 2000 in Pakistan. He took admission in The University of Lahore in Electrical Engineering degree program for a period of 4 years from 2019 – 2023. S.M Ibraheem is an active member of NSPE (National Society of Professional Engineers). He holds certificates from various prestigious universities of the world including Australian National University, University of Colorado, etc. During his first year in the university, he designed a software based COVID 19 project. In second year, he designed a quadcopter and AC-DC power supply project.

Book:

The purpose of the book is to provide the basic information on the aerial robotics and how the basic quadcopter is designed using STM32F100RB microcontroller. What are the basic mathematical equations, transformations and how a quadcopter flies in the air. In the book the basic algorithm, circuit and block diagram is explained.

After studying this book, the reader will be able to design, understand and explain the basics of the quadcopter and aerial robotics.

Table of Contents

Abstract ………………………………………………………………..7

Chapter 1: Introduction to Quadcopter

- Robotics…………………………………………………9 - 10
- Aerial Robotics……………………………………….10 - 11
- Quadcopter……………………………………………12 - 13
- Quadcopter benefits……………………………………..14

Chapter 2: STM32F100RB

- Introduction……………………………………………….16
- Cortex M3 Processor…………………………………….17
- Microcontroller Features……………………………….18

Chapter 3: Structure & Deign of Quadcopter

- Frame & Designing…………………………….20 - 21
- Landing Gear……………………………………..21 - 22
- Propellers & Motors…………………………….22 - 24
- Battery……………………………………………………24
- Electronic Speed Controllers…………………….25
- Flight Controller………………………………………25

Chapter 4: Mathematical Model of Quadcopter

- ➢ Frame of References……………………….27 - 28
- ➢ Linear Transformation…………………….29
- ➢ Mathematical Equations………………….30 - 34

Chapter 5: Flight Mechanics

- ➢ Six Degree Motion……………………………..37-38
- ➢ Throttle / Lift……………………………………….38
- ➢ Aileron / Roll……………………………………….39
- ➢ Rudder / Yaw………………………………….40 -41
- ➢ Elevator / Pitch…………………………………….41

Chapter 6: Circuitry of Quadcopter

- ➢ STM32 Flight Controller………………..43-45
- ➢ Block Diagram……………………………..46 - 47
- ➢ Pin Configurations………………………48 -49
- ➢ PDB Board…………………………………………..49
- ➢ MPU6050………………………………………49 – 50

Chapter 7: Programs and Algorithms

- ➢ Introduction……………………………………………52
- ➢ Algorithms……………………………………..52 - 56

Standard Definitions……………………………………..57

Abstract

The quadcopter is an electronics, mechanics and aerodynamics robot. The project is designed using STM32F100RB microcontroller. The quadcopter also known as UAV's which have high controllability and flight mechanics. The purpose of this book is to explain the design of a quadcopter During the design various software like Keil Micro-Vision 4, Auto Desk AutoCAD 2020, Proteus 8 and NI Multisim were used.

Chapter 1
Introduction To Quadcopters

Robotics

Robotics or mechatronics is the engineering field. This is a new type of engineering field. The field includes computer science, electronics engineering, artificial intelligence and mechanical engineering.

Figure 1 Humanoid Robot

In this figure a humanoid robot (a robot that acts and move like humans) is shown. These are advance and complex engineering machines.

Robots are taking a new place in our society they can perform tasks which are difficult for humans to perform. For example, mountain photography, handling of heavy material in the industry and etc.

The robotics is further divided into various fields like operator interface, programming, locomotives, sensing & perceptions and manipulators.

Aerial Robotics:

Aerial robots are also known as drones or **UAV's** (unmanned aerial vehicles). These robots are much advance robots and have a strong programming. These kinds of robots or drones covers aeronautical engineering, mechanical engineering, electronics engineering and programming concepts and knowledge.

Both the aircraft and the drone can fly in the air. But the major difference between them is that an aircraft is controlled by a human being whereas a drone is remotely controlled by a human thus a normal aircraft has a pilot in its cockpit whereas a drone has no pilot.

These drones are small in size and weigh only few hundreds of grams. Thus, these drones can go to the places where humans cannot go and can explore things which humans cannot do properly.

The aerial photography which was done by helicopters in the past is now done by drones. These drones are also capable to supply small amount load from one place to another place. A perfectly programmed and designed drone can fly high and can last longer in the air for about several hours.

Quadcopter

In the previous topic we introduced you to the robotics and the aerial robotics. The purpose here is to make you familiarize about quadcopter and how to build a fully working quadcopter.

Quadcopter is an aerial robot that has four propellers. A quadcopter is also known as a quadrotor, drone UAV. A quadcopter with three, five, six or eight propellers are called **tri copter, Penta copter**, **hexa copter**, and octa copter respectively. In the figure 2 we see a tri copter and it has three propellers while in figure 3 we have a hexa copter having 6 propellers.

Figure 2 Tri Copter   *Figure 3 Hexa Copter*

Aerial robotics includes different types of drones but a **quadcopter** is type of aerial robot that has four propellers and it moves by adjusting the speed of these propellers. In simple English a quadcopter is a helicopter that has four different propellers

Figure 4 Difference between a helicopter and a drone

whereas a normal helicopter has two propellers the main propeller and the tail propeller.

In the figure shown above we see a helicopter to the right and a quadcopter on the left. Here the helicopter has a main rotor whereas a small rotor is connected to tail of the helicopter to balance it, while in the case of quadcopter the quadcopter is fully controlled by its four propellers.

Quadcopter Benefits:

Some benefits of the quadcopter are discussed in the introduction. In this section we cover the **quadcopter benefits** in detail. These drones can make aerial photography and can help workers to inspect the construction site. These drones can also help in saving lives in time of crisis. These small drones are less noisy therefore, they can serve the purpose of security and surveillance. These drones can survey and inspect the dangerous sites. These drones can also be used in the agriculture to help farmers.

Chapter 2
INTRODUCTION TO STM32F100RB

Introduction:

STM32F100RBT6B is a microcontroller designed by *ST Microelectronics,* based on ARM CORTEX M3 microprocessor. This microcontroller comes in the discovery board series of the ST Microelectronics microcontroller series. This microcontroller is a low cost and high-performance motherboard, this quality makes it one of the best microcontrollers in the market to do embedded system projects.

Also, this micro controller is easy to program and understand, the 1000 pages long reference manual and data sheet provide enough information and guide lines for the beginner to work with the microcontroller.

Cortex M3 Processor:

The Cortex M3 is one of the first ARM based microprocessor. This processor which is based on ARMv7 – M architecture is designed to achieve high performance keeping the cost at minimum. The processor is based on Harvard Architecture, provides a 1.25 DMIPS/MHz of a benchmark score. Cortex M3 implements the debug technology in the hardware itself with several integrated components this technology provides the fast debug with trace and profiling.

Microcontroller Features:

As the microcontroller is based on the ARM cortex M3 processor it operates at 24MHz of the frequency with flash memory up to 128Kbytes and SRAM up to 8Kbytes. An extensive range of input / output peripherals are connected to the two APB buses. In microcontroller various communication interfaces are includes like one 12-bit ADC, two 12-bit DACs, up to six general-purpose 16-bit timers and an advanced-control PWM timer. The STM32F100xx low- and medium-density devices operate in the – 40 to + 85 °C and – 40 to + 105 °C temperature ranges, from a 2.0 to 3.6 V power supply.

Chapter 3
Structure & Design of quadcopter.

A quadcopter consists of many components. These components include

1. Frame
2. Landing Gear
3. Propellers & motors
4. Battery
5. Flight controllers
6. Electronics Speed controllers

Frame & Designing

A major component in quadcopter is its frame because all the motors, propellers flight controller battery rests on it. It is the **skeleton of the quadcopter.**

Frame can be designed by an **engineer** himself or an engineer can buy it from the market. These frames are available in variety of sizes and designs. Frame can be of aluminum, wood, plastic or **carbon fiber.** The frame can be selected on the basis of the requirement and the location. However, carbon fiber and aluminum frames are best for designing a quadcopter. The reason is that these frames are light in weight and have great strength.

Figure 5: X shaped Frame   *Figure 6: H shaped frame*

In figure 5 and 6 we have two different frames for the quadcopter. In figure 5 we have a **"X" shaped frame** whereas in the figure 6 we have a "H" shaped frame both frames are used according to the requirement. These frames are available in various sizes from small to large.

Landing Gear:

A landing gear is an **underneath area of an aircraft or a drone** and it is used for either takeoff or landing. It may be wheels or stand as shown in the figure below. In figure 7 we have a landing gear of an airplane whereas in figure 8 we have a landing gear of a drone or a small helicopter.

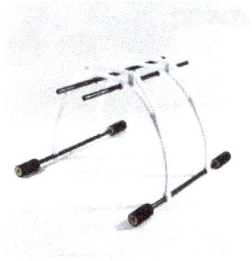

Figure 7 Aero plane Landing Gear

Figure 8: helicopter landing gear

Propellers and Motors:

Propellers and motors are collectively called engines of the quadcopter. A quadcopter has four propellers and four motors. In short, a quadcopter has four engines. These engines provide thrust (to push something forcefully in a specific direction) against the quadcopter and thus a quadcopter fly in the air.

The propellers and motors are available in different sizes according to the frames and requirement. The table 1 given below shows the propeller, motor, frame and battery sizes.

Table 1 Propeller, Frame, Motor and Battery sizes

Frame	Propeller	Motor	Battery
20 mm or smaller	3 inch	1104 – 1105	4000KV+
150 mm – 160 mm	3-4 inch	1306 – 1407	3000KV+
180 mm	4 inch	1806 – 2204	2600KV+
210 mm	5 inch	2204 – 2206	2300KV-2700KV
250 mm	6 inch	2204 – 2208	2000KV-2300KV
330 mm – 350 mm	7-8 inch	2208 – 2212	1500KV-1600KV
450 mm – 500 mm	9-10 inch	2212 – 2216	800KV-1000KV

The area that a propeller covers is called **the wing span** also known as the size of the propeller, some propellers have 2 wings, 3 wings or maybe 4 wings depending on the requirement.

Figure 9   Figure 10   Figure 11

Figure 12

Quadcopter or drone motors are different from the normal DC motors. The reason is that a quadcopter needs **high RPMs** (rotation per minute) and less weight motors. These motors are specially designed motors. They maybe out runner or in runner motors.

Battery:

A quadcopter must me light in weight in order to fly high and longer in the air. So normal batteries are not efficient for this purpose. Normally a quadcopter uses a 12V LiPo battery which is available in different mAh sizes.

The battery sizes are given in table 1.

Electronic Speed Controllers

Electronic Speed controllers are also known as ESC's. These are **small programable circuit**. An esc controls motor speed. The red, blue and yellow wires are connected to the motors. Other red and black wire is used to power up esc and a small controller wire is connected to the flight controller circuit which programs and instructs the esc. All four motors of the quadcopter have four individual escs.

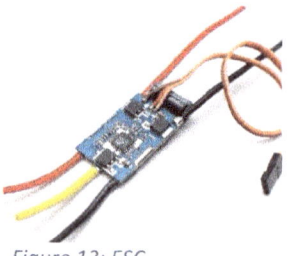

Figure 13: ESC

Flight Controller

A flight controller is the heart of the quadcopter or a drone. A flight controller can be designed using microcontrollers like Arduino, raspberry pie, **STM 32** and etc.

A flight controller gives and receives instructions from the remote control to the motor and controls the flight maneuvers.

Chapter 4
Mathematical Model of Quadcopter

A flight of a quadcopter is not easy to achieve, it is based on complex mathematical equations. The angular velocities, linear velocities, torque and force all are kept in mind while designing an amazing quadcopter. To find the all the

Frame of References:

An inertial frame of reference is an earth-based frame of reference which is fixed with respect to the earth whereas the body frame also known as vehicle frame of reference is fixed with the vehicle in our case fixed with the quadcopter.

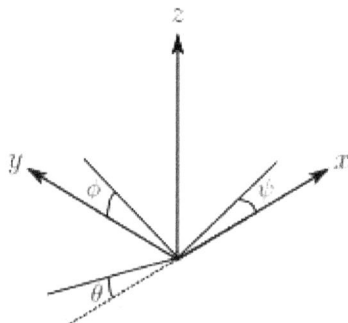

Figure 2 Inertial Frame of reference

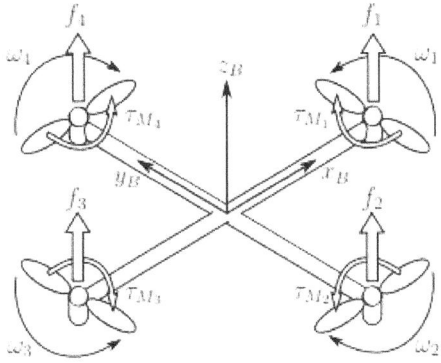

Figure 3 Body Frame

In inertial frame of reference

1. X – axis points toward the north / south
2. Y – axis points toward east / west
3. Z – axis points toward the earth / center of gravity.

In Body frame of reference

1. X – axis is aligned with the front motor
2. Y – axis is aligned with the right motor
3. Z – axis is cross product of the x any y axis

Linear Transformation:

In aeronautics either it is an airplane or space shuttle or a quadcopter all have the same matrix transformation. This is because a quadcopter or any flying machine performs rotation along x, y and z axis in inertial frame of references. Whereas these motions in body frame of references are called yaw, pitch and roll axis.

Actually, a quadcopter performs all of these rotation at once and does not perform individually. To perform these orientations the matrix of inertial frame of reference is transformed to body frame of reference using matrixes and transformation

Motion along X – axis is called pitch, its angle of rotation along Y -axis is called roll and the motion along Z – axis is called the yaw motion.

Mathematical Equations:

Let's denote that the inertial matrix is given by **α** which has x, y and z coordinates

$$a = \begin{bmatrix} x \\ y \\ z \end{bmatrix}$$

Similarly, the angles for these coordinates are **β** these angles are also known as Euler angles. Here pitch angle is **θ,** yaw angle is **ψ** and the roll angle is **φ**.

$$\beta = \begin{bmatrix} \phi \\ \theta \\ \psi \end{bmatrix}$$

A rotation matrix is needed in order to perform the transformation from the inertial frame of reference to the body frame of reference. Let's denote this rotation matrix by **γ** (gamma symbol)

$$R = \begin{bmatrix} \alpha \\ \beta \end{bmatrix}$$

Now the center of the body frame lies in the center of the gravity of the quadcopter. In body frame the linear velocities are determined by matrix V_L

And the angular velocities can be determined by V_A as shown in the matrixes given below

$$V_A = \begin{bmatrix} V_{xA} \\ V_{yA} \\ V_{zA} \end{bmatrix} \qquad V_L = \begin{bmatrix} V_{xL} \\ V_{yL} \\ V_{zL} \end{bmatrix}$$

Now these above matrixes show the linear and angular velocities in x, y and z direction

According to the figure given below it the rotation of X1 to X2 at an angle of "si" the rotation matrix for this will be

$$\begin{bmatrix} x1 \\ y1 \end{bmatrix} = \begin{bmatrix} x2 \cos(\psi) & - y2\sin(\psi) \\ x2\sin(\psi) & + y2\cos(\psi) \end{bmatrix}$$

Now the rotation matrix will look like

$$R = \begin{bmatrix} x1 \\ y1 \end{bmatrix} = \begin{bmatrix} \cos(\psi) & - \sin(\psi) \\ \sin(\psi) & + \cos(\psi) \end{bmatrix}$$

$$R(\theta, \phi, \psi) = R(x, \phi)R(y, \theta)R(z, \psi)$$

$$R(x, \phi) = \begin{bmatrix} 1 & 0 & 0 \\ 0 & \cos(\phi) & -\sin(\theta) \\ 0 & \sin(\phi) & \cos(\phi) \end{bmatrix}$$

$$R(y, \theta) = \begin{bmatrix} \cos(\theta) & -\sin(\psi) & 0 \\ \sin(\psi) & \cos(\psi) & 0 \\ 0 & 0 & 1 \end{bmatrix}$$

$$R(z, \psi) = \begin{bmatrix} \cos(\psi) & -\sin(\psi) & 0 \\ \sin(\psi) & \cos(\psi) & 0 \\ 0 & 0 & 1 \end{bmatrix}$$

Lets define the rotation matrix R from the body frame to the inertial frame of reference

$$R = \begin{bmatrix} \cos(\psi)\cos(\theta) & \cos(\psi)\sin(\theta)\sin(\phi) - \sin(\psi)\cos(\phi) & \cos(\psi)\cos(\phi)\sin(\theta) + \sin(\psi)\sin(\phi) \\ \sin(\psi)\cos(\theta) & \sin(\psi)\sin(\theta)\sin(\phi) + \cos(\psi)\cos(\phi) & \cos(\phi)\sin(\psi)\sin(\theta) - \cos(\psi)\sin(\phi) \\ -\sin(\theta) & \cos(\theta)\sin(\phi) & \cos(\theta)\cos(\phi) \end{bmatrix}$$

The rotation matrix is orthogonal in which the inverse of rotation matrix is equal to its transpose. In this case it is the rotation from inertial frame of reference to the body frame of reference.

To transform the velocity matrix from inertia frame to the body let it be $\acute{\omega}$. Whereas the transformation from body frame to the inertial frame is given by $\acute{\omega}^T$ So if we take the inverse or transpose of matrix $\acute{\omega}$. It will provide us with the opposite transformation.

$$V_L = \beta\acute{\omega}$$

The above equation is for moving from inertial frame to body frame of reference

$$\beta = V_L\acute{\omega}^T$$

In order to move from the body frame to the inertial frame of reference the above equation will be used.

The matrixes of the above two equations ae given below

$$V_L = \dot{\beta}\acute{\omega} \leftrightarrow \begin{bmatrix} V_{xA} \\ V_{yA} \\ V_{zA} \end{bmatrix} = \begin{bmatrix} 1 & 0 & -\sin(\theta) \\ 0 & \cos(\phi) & \cos(\theta)\sin(\theta) \\ 0 & -\sin(\phi) & \cos(\phi)\cos(\theta) \end{bmatrix} \begin{bmatrix} \dot{\phi} \\ \dot{\theta} \\ \dot{\psi} \end{bmatrix}$$

$$\beta = V_L \dot{\omega}^T \leftrightarrow \begin{bmatrix} \dot{\phi} \\ \dot{\theta} \\ \dot{\psi} \end{bmatrix} =$$

$$\begin{bmatrix} 1 & \sin(\phi)Tan(\theta) & \cos(\phi)Tan(\theta) \\ 0 & \cos(\phi) & -\sin(\phi) \\ 0 & \frac{\sin(\phi)}{\cos(\theta)} & \cos(\phi)/\cos(\theta) \end{bmatrix} \begin{bmatrix} V_{xA} \\ V_{yA} \\ V_{zA} \end{bmatrix}$$

As the quadcopter is symmetric in structure hence the inertial matrix is a diagonal matrix in which $I_{xx} = I_{yy}$ so

$$I = \begin{bmatrix} I_{xx} & 0 & 0 \\ 0 & I_{yy} & 0 \\ 0 & 0 & I_{zz} \end{bmatrix}$$

The angular velocity of the rotator is denoted with **ꞷi** creates a force in the direction of rotation axis denoted with **fi**. Now these angular velocities and accelerations create a torque around the rotation axis

$$fi = k\text{ꞷi}$$
$$\tau M = b\text{ꞷi}^2 + IM\dot{\omega}^2$$

Here k is the lift constant whereas the drag constant is b and the moment of inertia of the rotor is IM

The combined forces of the rotor create the thrust T in the direction of the body of the z axis whereas the torque along respective body frame angles is given by

τθ, τψ, τφ

$$T = \sum_{i=1}^{4} fi = k \sum_{i=1}^{4} w_i^2$$

$$T^B = \begin{bmatrix} 0 \\ 0 \\ T \end{bmatrix}$$

$$\tau B = \begin{bmatrix} \tau\phi \\ \tau\theta \\ \tau\psi \end{bmatrix} = \begin{bmatrix} lk(-w_2^2 + w_4^2) \\ lk(-w_1^2 + w_3^2) \\ \sum_{i=1}^{4} TM_i \end{bmatrix}$$

Here l is the distance from the rotor to the center of gravity of the quadcopter

Chapter 5
Flight Mechanics

Six Degree Motion

A quadcopter has **six degree of freedom** it can move upward, downward, left right, rotate left and rotate right. These six degrees of the motion is controlled by the motion of the wings of the quadcopter.

A quadcopter motors move in two different directions. Two motors move in clockwise direction and two motors move in anticlockwise direction as shown in the figure below

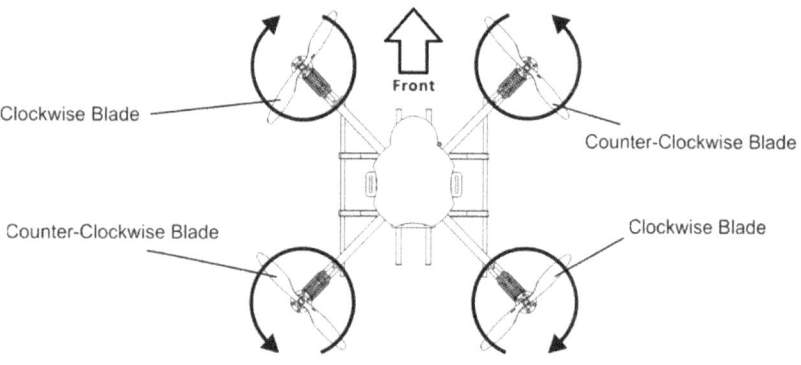

Figure 14

The rotation of motion of engines in counter and clockwise directions cancels the net torque produced by the quadcopter.

In six degree of motion a quadcopter has following motions

1. Aileron / Roll
2. Rudder / Yaw
3. Elevator / Pitch
4. Throttle / Lift

According to these motions the quadcopter moves up, down, left, right rotate left and rotate right

Throttle / Lift

To lift the quadcopter up. All the four engines must rotate at full speed equally. This means that engines will rotate at the same RPMS.

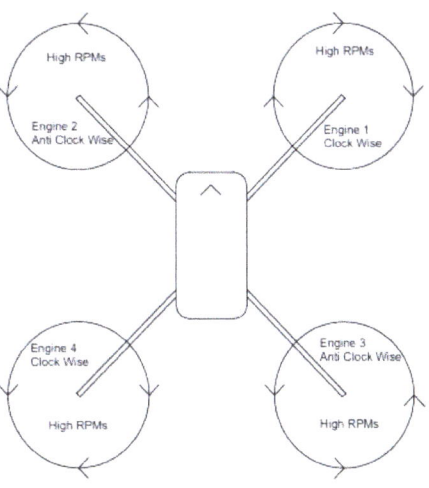

Figure 15: Lift Motion

Aileron / Roll

In aileron or the roll, the quadcopter will move left direction or the right direction. To rotate a quadcopter to the left direction the right two engines will move at higher RPMs then the left two engines.

To move the quadcopter to the right direction the left engines will move a high speed whereas the right two engines will move at a low speed

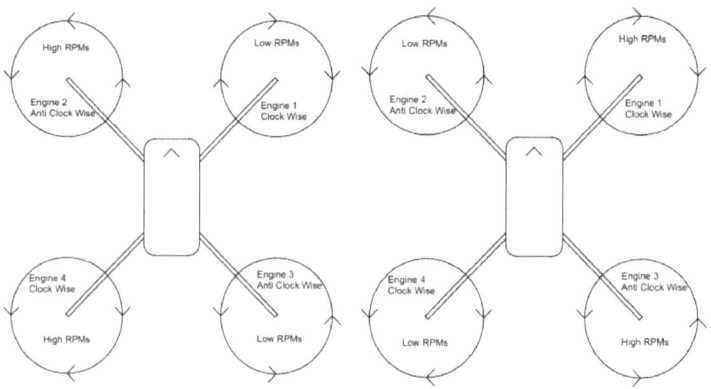

Figure 16: Roll Right   *Figure 17: Roll Left*

Rudder / Yaw

During the Yaw motion the quadcopter will either rotate left or right at the fix axis. In the case of roll motion quadcopter changed its axis with respect to the Earth. Whereas in Yaw motion the quadcopter will simple rotate 360 degree at left or right without changing any axis.

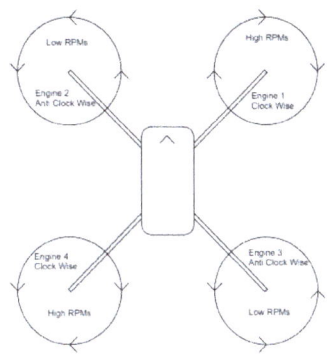

Figure 18: Yaw Right

To rotate left the anticlockwise engines will move at higher RPMs then the clockwise engines. This will create a torque for the quadcopter to rotate. This phenomenon will be accompanied by Newton third Law of motion

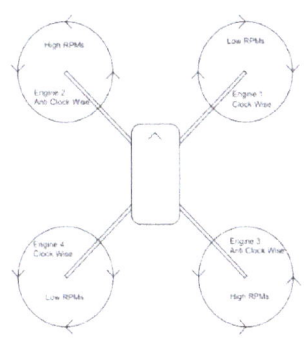

Figure 19: Yaw Left

To rotate right the clockwise engines will move at higher RPMs then the anticlockwise direction, generating enough torque for the quadcopter to rotate right direction

Elevator/ Pitch

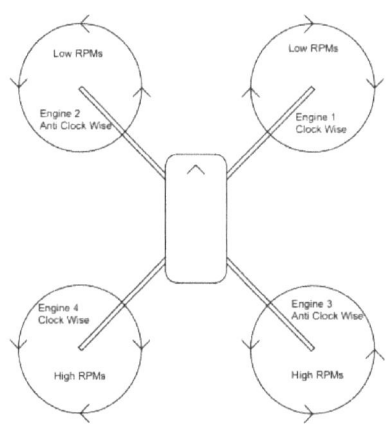

Figure 40: Pitch Forwards

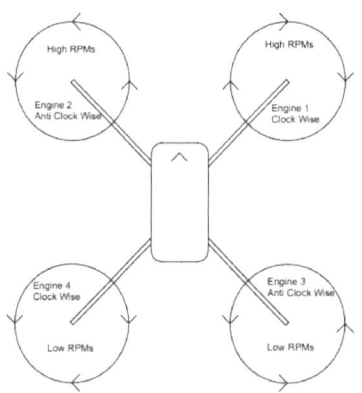

Figure 21: Pitch Backwards

Pitch of a quadcopter is referred to the forward and backward motion of the quadcopter. To move forward the rear two engines moves at higher RPMs whereas the front two engines spin at low RPMs. This produces more thrust at the rear side which pushes the quadcopter forward. Similarly, to move backward the front two engines move at higher RPMS than the rear to engines, creating high trust force at the front side which pushes the quadcopter to the back.

Chapter 6
Circuitry of Quadcopter

In this chapter the circuit of the quadcopter will be discussed. This chapter will cover the following circuit.

1. The Circuit of the STM32F100RB based Flight Controller
2. Block Diagram of the Quadcopter
3. Pin configurations of the STM32

STM32 Flight Controller

Mounting STM 32 on the quadcopter is not easy so a base circuit is designed to fit the STM32 on the quadcopter This also the flight controller of the quadcopter. The brain of the instruction. But some other circuit complete the flight controller circuit including RC transmitter's receiver, MPU-60-50, PDB board. But for now, let's discuss the STM32 Base circuit

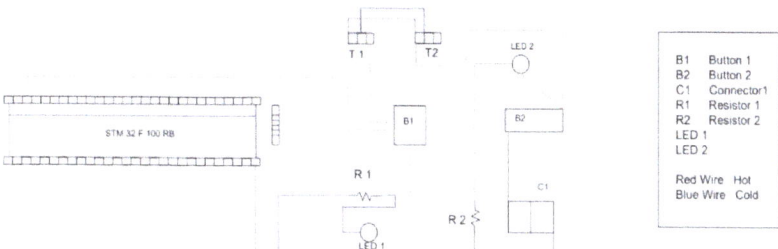

Figure 22: Circuit Diagram of the Flight Controller Base

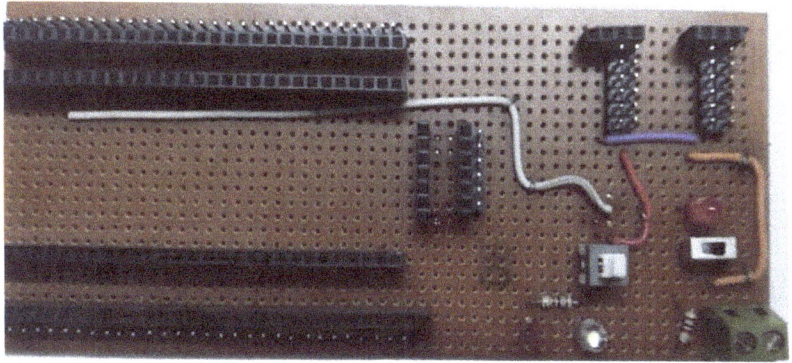

Figure 23: Flight Controller Base PCB

Figure 24: Flight Controller Base PCB Soldering

Figure 25: Flight Controller Base PCB with STM32 Mounted

Figure 24: Flight Controller Working

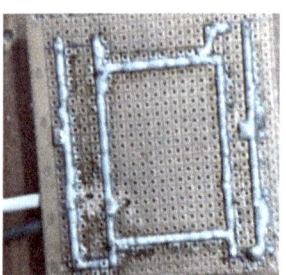

Figure 28: PDB Soldering

Figure 27: PDB Front

Block Diagram

Block diagram is the non-circuit representation of how might a circuit looks like or how the components will be connected.

According to the quadcopter's block diagram the engines motors will be connected to the ESC's and the ESC's will be connected to the flight controller and are powered by the PDB board which is connected to the battery

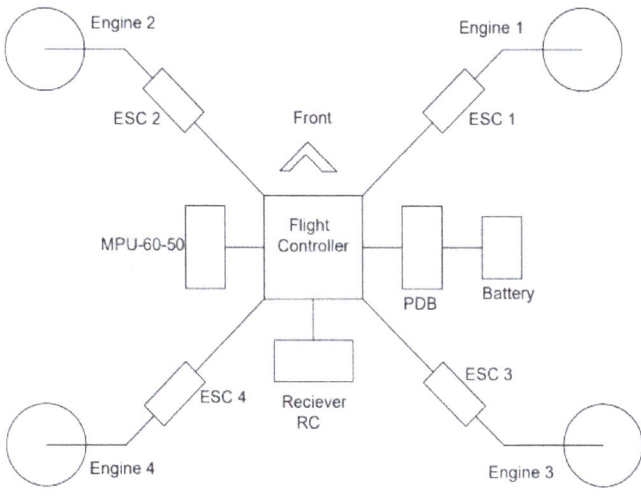

Figure 28: Quadcopter Block Diagram

The other components like MPU-60-50, RC transmitter receiver will be connected to the flight controller. The figure 28, tells us how these components will be connected to each other on the quadcopter. The following figure shows how the entire quadcopter is built.

Figure 28: Quadcopter Hardware build

Pin Configuration

The following tables tells us how the STM32F100RB pins are connected to the ESC's RC transmitter receiver and the MPU-60-50

STM 32	RECIEVER
PA0	CH 1
PA1	CH 2
PA2	CH 3
PA3	CH 4
PA6	CH 5
PA7	CH6
5V	BATT

STM 32	MPU – 60 -50
PB10	SCL
PB11	SDA
5V	VCC
GND	GND

STM32	ESC
PB6	ESC 1
PB7	ESC 2
PB8	ESC 3
PB9	ESC 4

Any type of transmitter / receiver controller can be used like Fly Sky transmitter and receiver

Power Distribution Board

Power Distribution Board or PDB is a special type of circuit board, whose purpose is to supply power to the quadcopter components like motors, camera, receivers etc from the battery. This PDB board is shown in figure 27 and 28. An engineer is responsible for the designs for PDB for the design project. It may vary for different projects.

MPU6050

MPU6050 is an MEMS (Micro Electro Mechanical System). It is one of the world first motion tracking device. It is designed for high performance while keeping the cost and power low. It has 3 axis of gyroscope and accelerometer which help in measuring velocity, orientation, acceleration and other motions

Figure 29 MPU60 50

It has on board Digital Motion Processor (DMP). DMP is responsible for solving complex mathematical equations. MPU6050 has in built 16 bit ADC (Analogue to Digital Converter). Due to ADC MPU6050 captures the 3D motion at the same time

Chapter 7

Programs & Algorithm

Introduction

The algorithm for the quadcopter has been designed using arbitrary values in C++ language

Rotor motion speed has been configured in percentage from 0 to 100%

But the standard chosen are 25, 50 and 75% for simple understanding

1,2,3,4 corresponds to the motor / esc 1, 2, 3,4 and the algorithm has been designed according to it

Algorithm

```
// Drone algorithm draft.cpp: This file contains the
'main' function. Program execution begins and ends
there.
//
// The algorithm of the program is written in C++
language
///////////////////////////////////////////////////
///////////////////////////////////////////////////
///

/*
      Algorithm is designed on the basis of activation
4 engines / rotors of the aerial robot

      As a quadcopter moves in a six degree of motion
axis

         Moves UP and DOWN                      LIFT
/ pitch
         Rotate LEFT and RIGHT                   YAW
         move LEFT and Right
ROLL
```

```
        A general RC transmitter has 6 different
channels with two joysticks
        The right joystick will be used to move
quadrotor up, down, left and right
        while the left joy stick will be used for
rotation of the quadcopter and lift the quadcopter

        STM 32 has various GPIO ports and pins
        Port A pins will be used to get data from the
Receiver of the RC transmitter
        Port B pins will be used to send the generated
the signal to the escs and the out-runner motors
        TO balance the quadcopter the gyro and accelro
of MPU6050 will be used

*/

void register_initlizations()
{
        // This function will declared to call /
initlize all the pins, ports, registers, of the STM 32
}
void MPU6050()
{
        // all the gyro and accelro coding will be done
here and will be called in the main function
accordingly
}

void receiver()
{
        // here all the receiver values will be called
from the port a
        int GPIOA_PIN0, GPIOA_PIN1, GPIOA_PIN2,
GPIOA_PIN3, GPIOA_PIN4, GPIOA_PIN5;
        int channel1, channel2, channel3, channel4,
channel5, channel6;


        GPIOA_PIN0 = channel1;
                // input from the IDR register 1
```

```c
        GPIOA_PIN1 = channel2;
        // input from the IDR register 2
        GPIOA_PIN2 = channel3;
        // input from the IDR register 3
        GPIOA_PIN3 = channel4;
        // input from the IDR register 4
        GPIOA_PIN4 = channel5;
        // input from the IDR register 5
        GPIOA_PIN5 = channel6;
        // input from the IDR register 6;
}
int main()
{
        // in the main function all the motors will be
powered from the battery and the receiver signals
        // PWM()
        MPU6050(); // mpu library called for the
balancing  of the quadcopter before flying
        receiver(); // all the receiver values will be
called
        int GPIOB_PIN0, GPIOB_PIN1, GPIOB_PIN2,
GPIOB_PIN3, GPIOB_PIN4, GPIOB_PIN5;
        int esc_1, esc_2, esc_3, esc_4;
        int left_joystick, right_joystick;

        // Lift the quadcopter
        // to move a joystick up dowm left right assume
values 1,2,3,4 respectively
        if (left_joystick == 1)
        {
                // throttle high
                int lift; // == 0 - 100%
                esc_1 = GPIOB_PIN0 = lift;
                esc_2 = GPIOB_PIN1 = lift;
                esc_3 = GPIOB_PIN2 = lift;
                esc_4 = GPIOB_PIN3 = lift;
                // all the motor will move at the same
speed in lifting the quadcopter
        }
        // hovering
        if (left_joystick == 1)
        {
```

```c
            // throttle high
            int lift = 50; // lift is permanently set to 50% for the hovering
            esc_1 = GPIOB_PIN0 = lift;
            esc_2 = GPIOB_PIN1 = lift;
            esc_3 = GPIOB_PIN2 = lift;
            esc_4 = GPIOB_PIN3 = lift;
            // all the motor will move at the same speed in lifting the quadcopter
    }
    // pitch
    if (right_joystick == 1)
    {
            esc_1 = GPIOB_PIN0 = 25;
            esc_2 = GPIOB_PIN1 = 75;
            esc_3 = GPIOB_PIN2 = 25;
            esc_4 = GPIOB_PIN3 = 75;
    }
    else if (right_joystick == 2)
    {
            esc_1 = GPIOB_PIN0 = 75;
            esc_2 = GPIOB_PIN1 = 25;
            esc_3 = GPIOB_PIN2 = 75;
            esc_4 = GPIOB_PIN3 = 25;
    }
    //yaw
    if (left_joystick == 1)
    {
            // throttle high
            int lift = 50; // lift is permanently set to 50% for the hovering
            esc_1 = GPIOB_PIN0 = lift;
            esc_2 = GPIOB_PIN1 = lift;
            esc_3 = GPIOB_PIN2 = lift;
            esc_4 = GPIOB_PIN3 = lift;
            // all the motor will move at the same speed in lifting the quadcopter
    }
    // roll
    if (right_joystick == 3)
    {
            esc_1 = GPIOB_PIN0 = 75;
```

```c
        esc_2 = GPIOB_PIN1 = 75;
        esc_3 = GPIOB_PIN2 = 25;
        esc_4 = GPIOB_PIN3 = 25;
    }
    else if (right_joystick == 4)
    {
        esc_1 = GPIOB_PIN0 = 25;
        esc_2 = GPIOB_PIN1 = 25;
        esc_3 = GPIOB_PIN2 = 75;
        esc_4 = GPIOB_PIN3 = 75;
    }
    //yaw
    if (left_joystick == 3)
    {
        esc_1 = GPIOB_PIN0 = 25;
        esc_2 = GPIOB_PIN1 = 75;
        esc_3 = GPIOB_PIN2 = 75;
        esc_4 = GPIOB_PIN3 = 25;
    }
    else if (left_joystick == 4)
    {
        esc_1 = GPIOB_PIN0 = 75;
        esc_2 = GPIOB_PIN1 = 25;
        esc_3 = GPIOB_PIN2 = 25;
        esc_4 = GPIOB_PIN3 = 75;
    }


}
```

Standard Definitions

Aerial Robotics:

An aerial robot is a system capable of sustained flight with no direct human control and able to perform a specific task.

Flight Controller:

The flight controller is the brain of the aircraft. It's a circuit board with a range of sensors that detect movement of the drone

Microcontroller:

A microcontroller is a small computer on a single metal-oxide-semiconductor integrated circuit chip. A microcontroller contains one or more CPUs along with memory and programmable input/output peripherals.

Microprocessor:

A microprocessor is an electronic component that is used by a computer to do its work. It is a central processing unit on a single integrated circuit chip containing millions of very small components including transistors, resistors, and diodes that work together.

THE END

www.ingramcontent.com/pod-product-compliance
Lightning Source LLC
Chambersburg PA
CBHW070814220526
45466CB00002B/658